Ducks

by Linda Carroll
illustrated by Eileen Hine

HMH

Copyright © by Houghton Mifflin Harcourt Publishing Company

All rights reserved. No part of this work may be reproduced or transmitted in any form or by any means, electronic or mechanical, including photocopying or recording, or by any information storage and retrieval system, without the prior written permission of the copyright owner unless such copying is expressly permitted by federal copyright law. Requests for permission to make copies of any part of the work should be submitted through our Permissions website at https://customercare.hmhco.com/contactus/Permissions.html or mailed to Houghton Mifflin Harcourt Publishing Company, Attn: Intellectual Property Licensing, 9400 Southpark Center Loop, Orlando, Florida 32819-8647.

Printed in the U.S.A.

ISBN 978-1-328-77225-1

4 5 6 7 8 9 10 2562 25 24 23 22 21

4500844736   A B C D E F G

If you have received these materials as examination copies free of charge, Houghton Mifflin Harcourt Publishing Company retains title to the materials and they may not be resold. Resale of examination copies is strictly prohibited.

Possession of this publication in print format does not entitle users to convert this publication, or any portion of it, into electronic format.

The ducks want to go for a swim.
How will they get there?
Hooray! Here is a truck!

2   How many ducks want to go for a swim?

One more duck wants to swim.

The little duck runs to the truck.

One more duck gets in.

How many ducks are in the truck?

The ducks in the truck see a pond.
Hooray!
There are ducks in the pond.

How many more ducks are in the truck than are in the pond?

4

The ducks get out of the truck.
They go to the pond. They swim.

How many more white ducks than brown ducks?

Two more ducks want to swim.
They run, run, run, and jump in.

How many fewer brown ducks than white ducks?

One more duck comes.
Then they all fly away.
It has been a ducky day. Hooray!

How many ducks fly away?

Responding

Vocabulary

Duck!

Draw

Look at page 5. On a piece of paper, draw the ducks you see in the pond. Use an X to show each duck.

Tell About

Note Important Details Look at page 5. Tell how many ducks are on the left and how many ducks are on the right. Tell how many ducks there are in all.

Write

Look at page 5. Write how many ducks there are in all.